Idia Techne
STEM Explorer

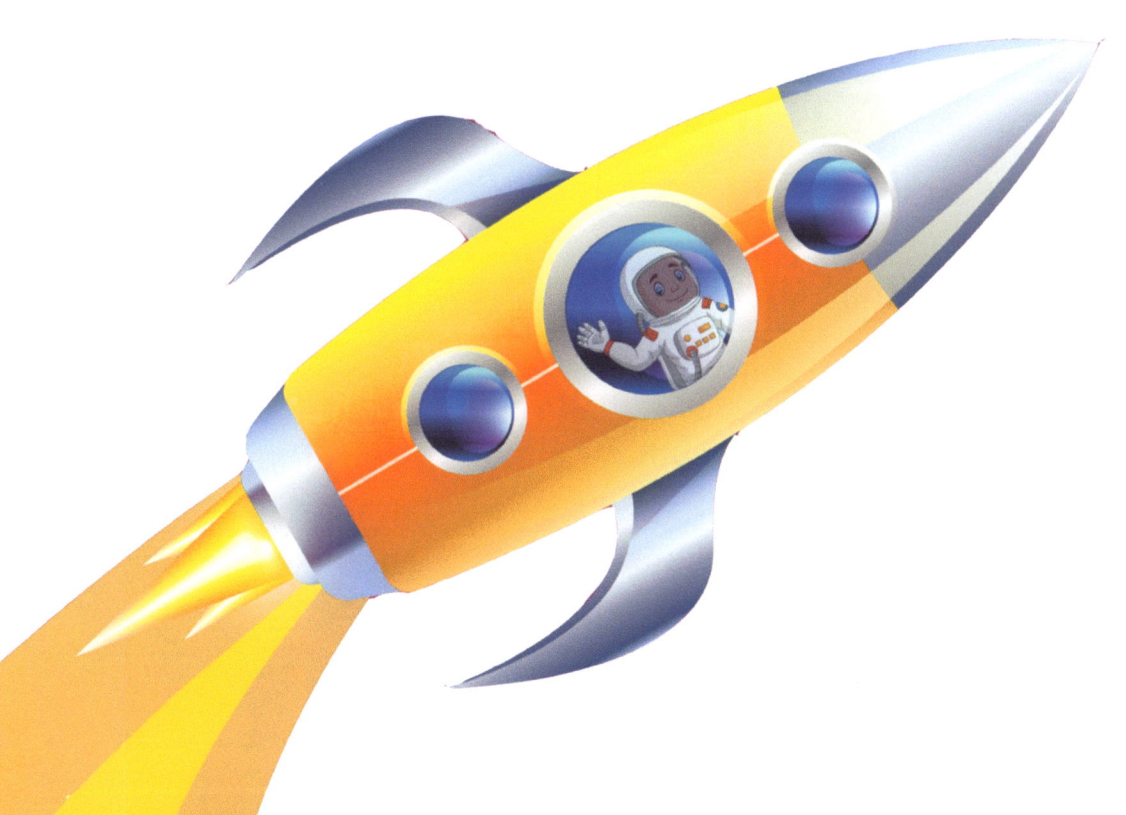

Iria Osara

Idia Techne - STEM Explorer © Iria Osara 2024

ISBN 979-8-9908900-0-8
Published by TYT TechED LLC 2024

Edited by Wrate's Editing Services

Library of Congress Cataloguing-in-Publication Data is on file with the publisher.

PRINTED IN THE USA

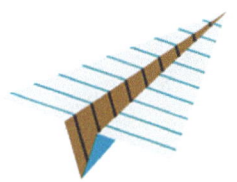

To AB, Raphael, Alexander and Oseiwe. You are my everything.

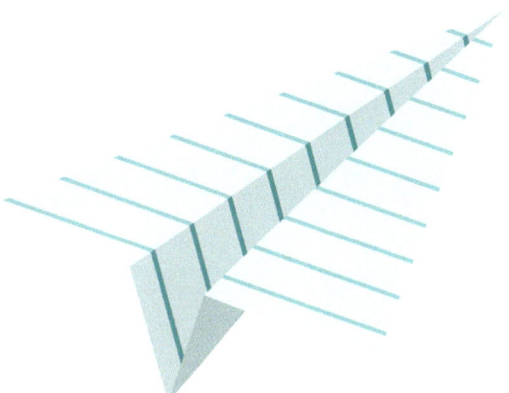

Hello, STEM adventurers!

I'm Idia Techne, your guide to the incredible STEM Galaxy.

Get ready for an exciting journey!

We're going to hop on a rocket and zoom off to some fun and fascinating planets.

We'll meet the friendly people who live there and discover what they do.

STEM is short for something amazing:

S stands for **Science**. It's the superpower planet that helps us to understand everything around us.

T stands for **Technology**. It's the magical planet where our dreams come alive.

E stands for **Engineering**. It's the construction planet where our imagination becomes a reality.

M stands for **Math**. It's the planet of shapes and numbers, and it's full of surprises.

So, buckle up, hold on tight, and let's blast off

into a world of adventure and learning.

The first planet on our way is Planet Science.

Science is a super cool planet. It helps us to understand everything around us.

The people who live there are called **Scientists**.

You might be asking:
    Why do we need Science?
    Who lives on Planet Science?
    What are the benefits of Science?

Don't you worry, as people *always* have lots of questions about Science.

Today, we'll explore how Science helps us to make the world a better place.

You probably know about doctors, nurses,
dentists, biologists, and chemists…
so, who else?

Let's go and find out.

Archeologists

They help us to discover our history and culture.
Things that happened millions of years ago.

How amazing is that?

Heard about **Paleontologists?**

Like **Archeologists**, they uncover past secrets…

They use a range of tools for their work, including brushes, measuring tapes, storage bags, pens, and pencils. And these helped them to discover the existence of…

…dinosaurs!

Lab Technicians

Lab Technicians are like detectives – they use special equipment to make discoveries.

Geologists

A **Geologist** is someone who studies rocks, soil, and the earth.

Thanks to Geologists, we can learn about our amazing Earth and use its gifts wisely!

They are our Earth Detectives.

Epidemiologists

An **Epidemiologist** is a scientist who helps to keep people healthy by studying how diseases spread and figuring out ways to stop them.

Dieticians

A **Dietician** is a superhero who helps us to choose the foods that make our bodies strong and healthy. They are determined to get you eating your fruit and veggies!

Agronomists

An **Agronomist** is a scientist who studies agriculture.

They possess all the secrets of how to protect plants from pests and diseases.

This helps our farmers to grow the food on your plate.

Oceanographers

An **Oceanographer** is a cool explorer who discovers the secrets of the big blue sea and all the amazing creatures that live in it, including seahorses, sharks, starfish, octopuses, and penguins!

These are just a few of the people who live on **Planet Science**.

We also have **Neuroscientists, Environment AI scientists, Geneticists, Meteorologists...**

... can you think of any more?

Planet Science was so much fun – being a scientist must be really cool.

Up ahead is **Planet Technology**.

Technology – the tie that binds all the planets together. It is a type of science that's used to solve problems and invent cool and useful machines.

A person who specializes in Technology is called a **Technologist**.

One of the most popular areas of Technology is called Computer Science.

Computer Scientists work in many different fields, including…

Software Development
Artificial Intelligence
Cybersecurity
Computer Networks
Robotics and Automation
Game Development

Cybersecurity

Game Development

Before we explore these fields further, let's meet some real-life **Computer Scientists**.

Robotics

Computer Networks

Grace Hopper was one of the first
Computer Programmers.

She made a special language called FLOW-MATIC
that made it easier for people to tell computers what
to do. People loved it so much they created another
language called COBOL, which is still used today,
more than 60 years on.

Gladys West is a female Mathematician who played a critical role in the creation of the global positioning system (GPS).

GPS is a clever computer system that works like a map. It can pinpoint where we are and tell us the best route to our destination. How would we ever find our way without Gladys's help?

AI Engineer

They use skills from Computer Science and Mathematics to create super smart machines that can learn and think just like us. Sounds like magic, doesn't it?

With all the cool things here on Planet Technology, who helps to spread the word about them?

Well, **Developer Advocates** are like the newscasters of Technology – they tell us everything that is going on.

And let's not forget the Support Team. Support cuts across every area of STEM.

Support Engineers help us to find the answers to our computer problems. Planet Technology has many superheroes, and support engineers are among them. They are the backbone of any organization.

With so much software and so many apps, how do people learn how to use them?

Technical Writers – they are the storytellers of Planet Technology.

They prepare the documents that guide us through learning new software or using a new app.

I love how everything on Planet Technology is connected. It's just like a spider's web! Let's meet the people who link everything together.

Network Engineers are IT specialists who design, build, and configure computer systems such as the Internet.

The Internet is often referred to as 'The Cloud'. But unlike the physical clouds in the sky, where rain forms, it's a virtual creation of computers that need maintaining by experts.

Cloud Architects are like builders. They create special computer houses in the sky that allow us to store and share information with our friends and family.

I saw someone checking to make sure I could access Planet Technology earlier. Who was that?

A **Cybersecurity Analyst** – they ensure our computer systems and networks are protected from theft or damage… just like the security guards at the mall do!

I like how Planet Technology is so easy to access. Who designed it this way?

A **UX Designer** – they create computer experiences that are easy and fun for everyone to use.

There's so much to see on Planet Technology,
but we have a third planet to visit –
Planet Engineering!

Engineering is a branch of Science and Technology.

Engineers help to invent, design, and build complex machines and systems such as airplanes, cars, ships, trains, and tractors.

I hear there are many different types of Engineers.

So, it will be nice to meet some of the people that have made Planet Engineering their home.

Let's go!

This place looks amazing and beautiful.
Who helped to create it?

A **Civil Engineer.** They build and maintain
our roads, bridges, buildings, and water
resources.

With all that construction going on, who helps to protect Planet Engineering from pollution?

Environmental Engineers! They are often referred to as the guardians of the Earth, as they find ways to reduce waste and contamination.

Wow! Now, I would love to see what all the planets look like from the sky.

Who are the superheroes that can make this happen?

Aerospace Engineers design things that fly, like planes and rockets.

I smell something yummy! Who's cooking?

Food Process Engineers are the kitchen maestros who use their knowledge to design and improve the methods used to make our favorite foods, from chocolate bars to potato chips.

That was a yummy meal. Who will help me to clean up so I can visit the final planet?

Oh, a robot! Nice. Who made it?

Robotics Engineers are like the wizards of Planet Engineering. They design and build robots that can do all sorts of things, from cleaning your house to exploring other planets.

Last but not least, we are here at **Planet Math.**

Math is short for Mathematics, which is the study of numbers, shapes, and formulas.

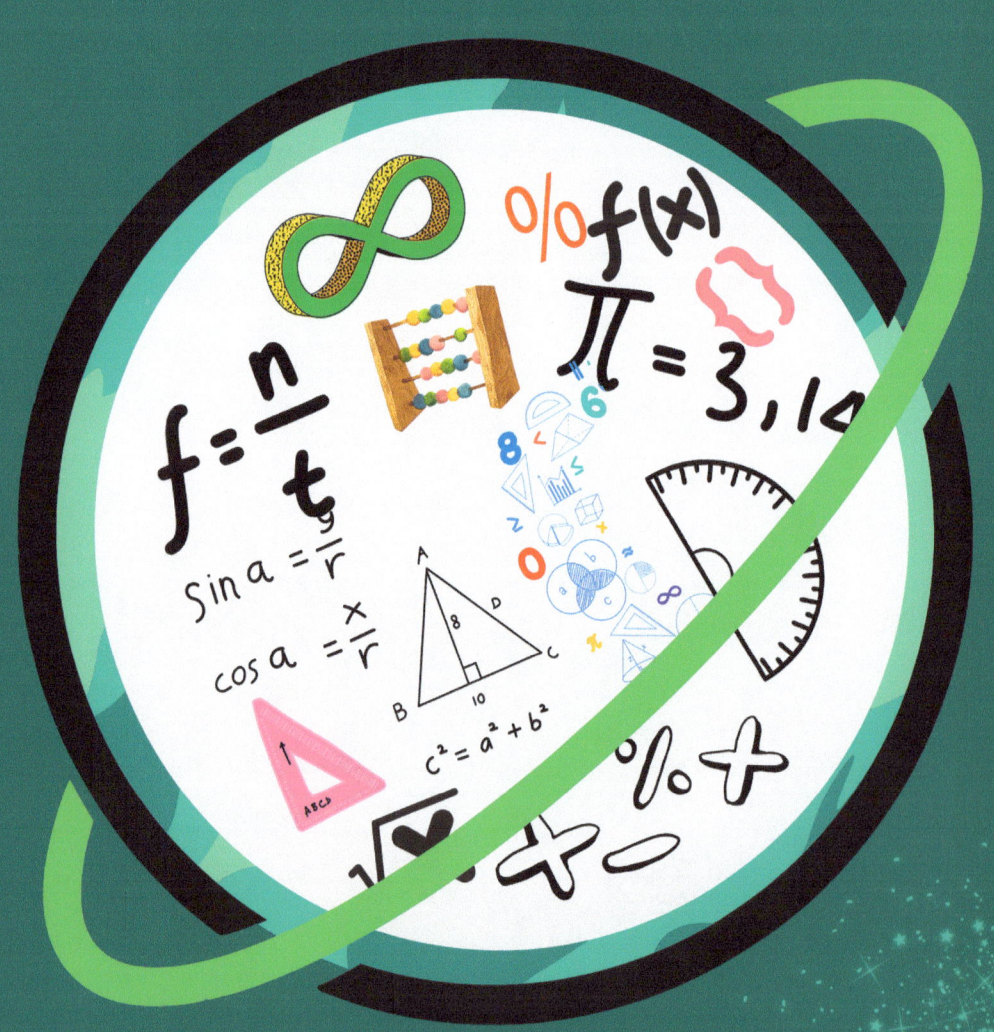

It does not relate to STEM alone –
even artists use Math in their work!

Math is like a language that helps us to understand Science, Technology, Engineering, and even Art.

Without Math, we wouldn't know the right amount of flour and eggs to use for baking our cookies, what shapes to cut them in or how to count the money in our piggy bank.

Wow! So, Math is like a friendship chain holding everything together.

Let's meet some real **Mathematicians**...

Many, many years ago, sailors needed a way to find their way across the seas – they weren't lucky enough to have GPS or phones like we do today.

Instead, they looked up to the skies to determine their location…

Mathematician **Hypatia of Alexandria** developed a system for sailors to navigate the world using the stars. Talk about having star quality!

Just like we have locks to keep our important things safe, cryptography in Math helps to protect our online information.

A Mathematician named **Julia Robinson** did important work in this area to make sure our information stays secure.

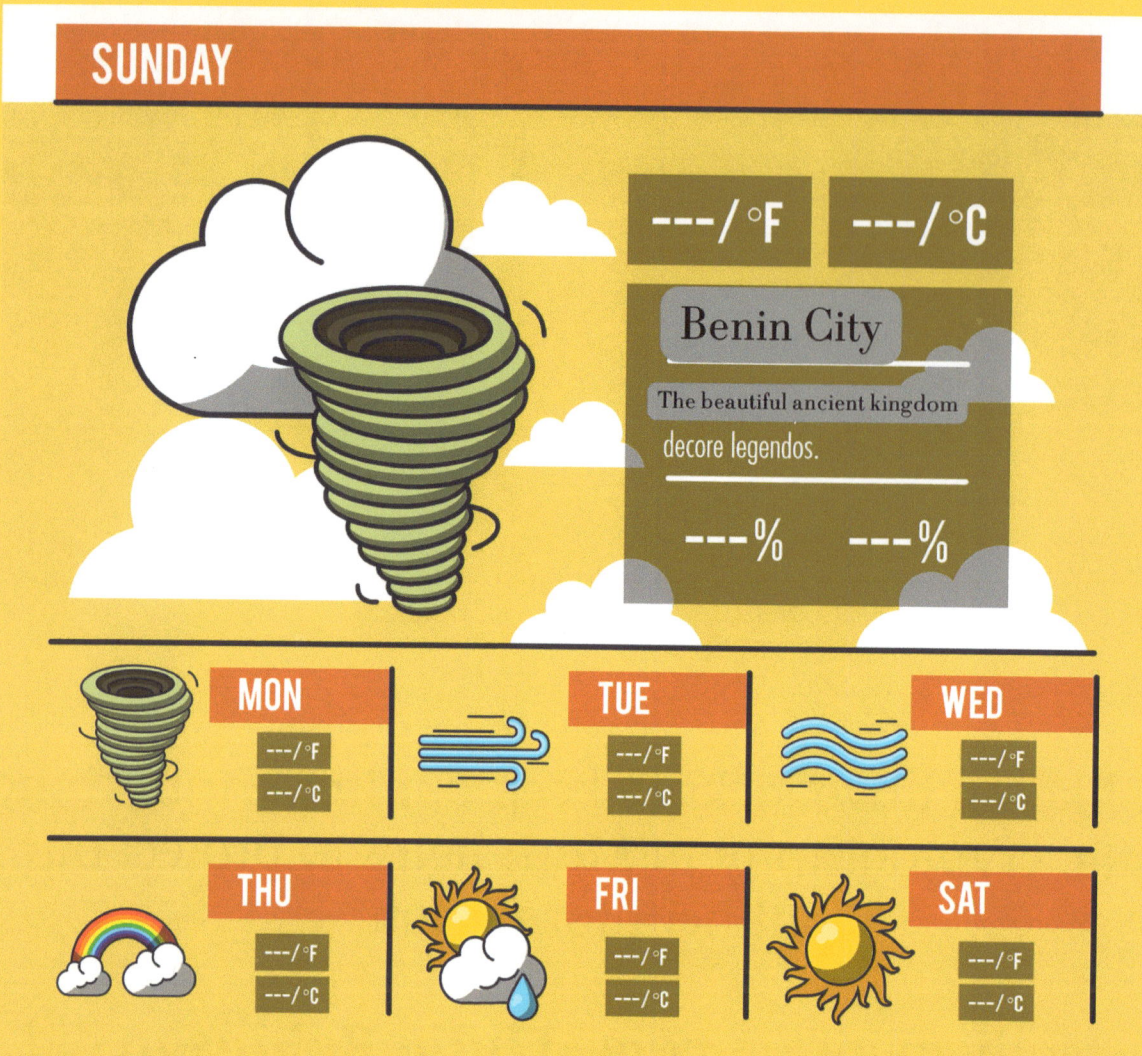

Back in the olden days, people didn't have the skills to predict the weather. Imagine going to the beach without knowing whether it was going to rain later.

Thanks to the British Mathematician **Mary Cartwright**, we can forecast the weather. She did this by using equations to figure out how air pressure, temperature, and wind affect the weather.

Sadly, we have now visited all the planets in the STEM Galaxy and our journey has come to an end. Let's continue exploring this amazing galaxy by doing some fun activities.

Create a Rainbow in a Jar

You will need:

»   A see-through jar
»   A jug of water
»   Food coloring
»   Vegetable oil
»   Dish soap

Instructions:

Step 1 – Pour the water into the jar, until it reaches approximately a third of the way.

Step 2 – Add a few drops of food coloring followed by several drops of dish soap.

Step 3 – Slowly pour in the vegetable oil.

Step 4 – Hold the jar up to the light and observe the amazing results.

Make a Binary Bracelet

Binary Code

Computers use a special language called binary code, which is made up of zeros and ones. It's called binary because it has only two options – just like how a light switch can be either on (1) or off (0).

With just a few materials, you can make your own personalized bracelet using binary codes.

You will need:

» Some colored beads (example red and white)

» String

Instructions:

Step 1 – Decide on a color to represent 0 and another for 1.

For example, red is 0 and white is 1.

Step 2 – Let us use the beads to spell out US.

Computer language for the letter U is 01010101

Computer language for the letter S is 01010011

Step 3 – Put U and S together: 01010101 01010011

Step 4 – Now you have a bead bracelet that says US in binary code!

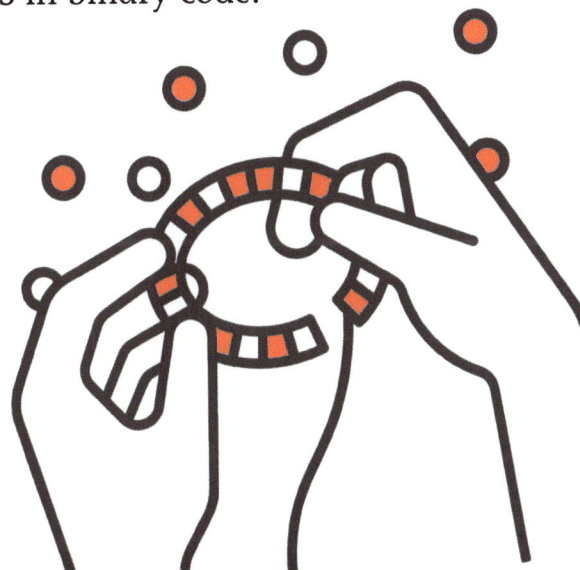

Fold a Paper Airplane

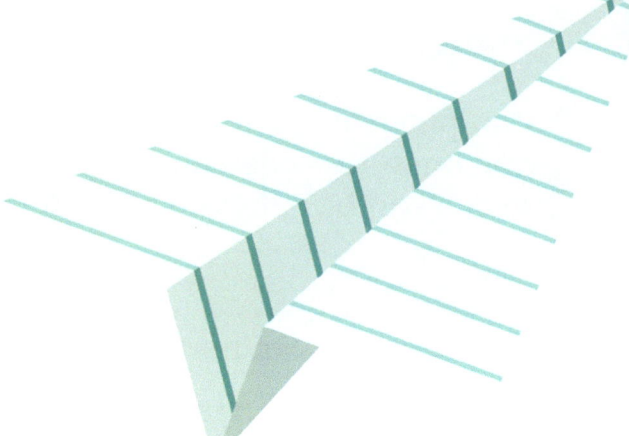

You will need:

- » A piece of paper, 8.5 x 11 inches in size
- » A ruler
- » A pencil
- » Scissors
- » A tape dispenser

Instructions:

Step 1 – Fold the paper in half lengthwise.

Step 2 – Unfold the paper, and then fold the top corners down to the center crease.

Step 3 – Next, fold the top edges down to the center crease.

Step 4 – Fold the paper airplane in half along the center crease.

Step 5 – Fold the wings down, aligning the top edges with the bottom edge of the body.

Step 6 – Fold the back of the plane upwards, creating a tail.

Step 7 – Secure the wings and tail in place with tape.

Congratulations! You have successfully made a paper airplane.

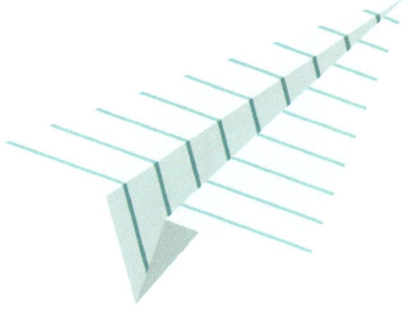

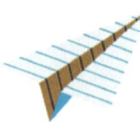

Throw a Pizza Party!

Here's a fun way to practice counting, recognize shapes and improve your math skills while pretending to enjoy some tasty pizza!

You will need:

- » A piece of paper
- » Markers
- » Scissors

Instructions:

Step 1 – Draw a big pizza on your paper and cut it out.

Step 2 – Divide your pizza into several slices, just like a real pizza. Have fun and make them any shape or size you like.

Step 3 – Give each slice of pizza a number from 1 to 5.

Step 4 – To make the game more fun, you can also draw toppings on each pizza slice.

Instructions for parents and carers:

1. Counting: Ask the child to count the number of pizza slices with you. Point to each slice and say the numbers out loud. Encourage them to count from 1 to 5 or count backward from 5 to 1.

2. Shape Identification: Show the child the different pizza slices and ask them to identify the shapes. For example, "Can you find the triangular slice?" or "Which slice has a curved shape?" This activity will help them to recognize and name different shapes.

3. Stacking: Have the child stack the pizza slices on top of each other. Ask them to count how many slices are in the stack. You can also ask questions like, "If we add one more slice, how many will there be?"

Glossary

Binary   A system used to communicate with a computer. It uses only two numbers: 0 and 1.

Computer   An electronic machine that helps us to do fun things like play games or watch movies.

Data   Information that we collect and use to learn about things. It can involve numbers, words, pictures, or even sounds.

Developer   A broad term used to refer to a Software Engineer. Software Engineers help to create applications and software (including games!).

Field   A specific area within a profession. For example, in Computer Science, we have Data Analysts and Software Engineers, etc.

Forecast   Involves predicting what might happen in the future.

Network   Similar to a big spider's web, it connects different computers and devices.

Programmer   Someone who uses codes to tell a computer what to do. These instructions are called programs.

Robot   A special machine that can do things by itself. It can move and talk.

Software   A set of programs (including games!).

Techne   A Greek word that means the use of knowledge, techniques, and skills to develop innovative tools or solutions.

UX   Short for User Experience.

Weather   What the air and sky are like outside. It tells us if it's sunny, cloudy, rainy, snowy, or windy.